## Table of Contents

# Wellspring of Prosperity
## Science and Technology in the U.S. Economy

How Investments in Discovery Are Making Our Lives Better

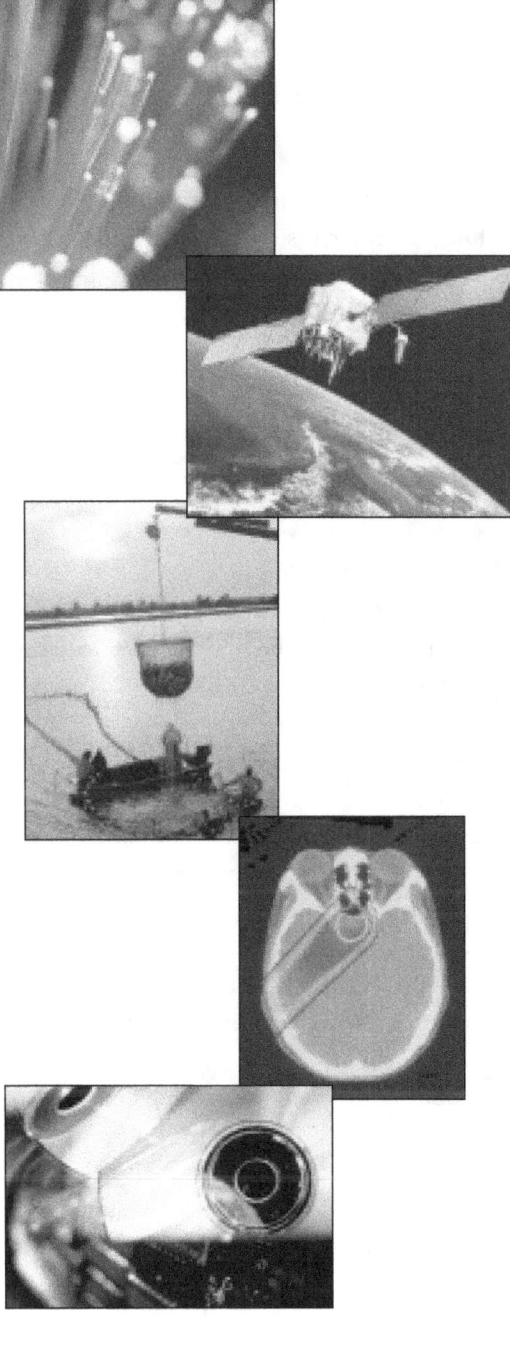

| | |
|---|---|
| **Foreword** | ii |
| **Message from the Vice President** | iv |
| **Introduction** | 1 |
| Chapter 1<br>**Information Technologies** | 3 |
| Chapter 2<br>**Global Positioning System** | 7 |
| Chapter 3<br>**Biomedical Technologies** | 11 |
| Chapter 4<br>**Food Technologies** | 15 |
| Chapter 5<br>**Environmental Technologies** | 19 |
| Chapter 6<br>**Manufacturing Technologies** | 23 |
| **Epilogue** | 29 |
| **About PCAST** | 30 |

Since the earliest days of our nation's birth, the American experience has been defined by our boundless search for new frontiers, our ceaseless quest for new discoveries, and our restless pursuit of new knowledge.

Today, perhaps more than at any time in our history, the strength of our economy, the health of our families, and the quality of our lives depend upon America's unmatched science and technology enterprise. At the dawn of a new century, much of our success in the years ahead will depend in large measure on the investments we make today in scientific research and technological innovation.

President Clinton and I are unequivocally committed to sustaining and nurturing U.S. scientific leadership across the frontiers of scientific knowledge. This is not merely a cultural tradition of our nation. It is an economic and security imperative. We must rise to this challenge while ensuring that our newest and most advanced technologies incorporate our oldest and most cherished values.

Whether measured in terms of discoveries, citations, or prizes, our country's prior investments have yielded a scientific and engineering enterprise without peer. Over the past several decades, public investments in research have helped America's scientists and engineers split the atom, splice the gene, explore the moon, invent the microchip, create the laser, and build the Internet — and in the process millions of good-paying jobs have been created.

To spur America's future achievements in science and engineering, the Administration has acted in a variety of roles: sustaining our research leadership position; strengthening a business environment that supports private sector research and development; investing in technological infrastructure; and advancing critical technologies, often in partnership with our universities and industries. The accomplishments and initiatives described in this report are representative of our research and development portfolio that has enjoyed broad bipartisan support.

It is, of course, impossible to accurately predict which areas of science and engineering will yield ground-breaking discoveries, what those inventions will be, how they will impact other scientific disciplines and, eventually, benefit our daily lives.

Who can be sure exactly what advances will be needed to maintain our national security and our strong economy, or clean up our environment and develop a healthier, better-educated citizenry?

What we can ensure is that America remains at the forefront of scientific capability by sustaining our investments in basic research, thereby enhancing our ability to shape a more prosperous future for ourselves, our children, and future generations while building a better America for the twenty-first century.

Al Gore

# Introduction

# Wellspring of Prosperity
## Science and Technology in the U.S. Economy

As experts seek explanations for America's record-breaking performance in the current world economy, it is tempting to credit our bountiful natural resources and our diverse, hardworking population. But many other countries — and even our own ancestors — have enjoyed similar resources without producing an economic boom such as the United States achieved in the late 20th century. The credit for our recent success really goes to the powerful system we have generated to create new knowledge and develop it into technologies that drive our economy, guarantee our national security, and improve our health and quality of life.

This report explores some of the remarkable, tangible benefits to our nation from the sustained funding of research and development (R&D). The President's Committee of Advisors on Science and Technology (PCAST) prepared this report in response to Vice President Gore's request. The Vice President called for a new compact between our scientific community and our government — one that would emphasize support for fundamental science and a shared responsibility to shape our breakthroughs into progress. The Vice President turned to PCAST for help in illuminating the benefits of our past and present investments in science and technology.

The technologies covered here — those dealing with information, global positioning, biotechnology, food, the environment, and manufacturing — illustrate how much we have come to rely on technology and take it for granted in our everyday lives. But the marvels of today are really the fruits of research seeds planted decades ago — investments that have not only given us new technologies, but have also helped educate generations of engineers and scientists who now form an essential component of our modern workforce. The very fact that these advances required decades of investment stands as a warning against complacency in our future investment strategy. As the Vice President has noted, the government and the private sector must work together to ensure that today's investments in research and development are sufficient to yield similar payoffs to society in the 21st century. Close cooperation with our international partners is also crucial to the success of this venture, since scientific breakthroughs occur with no regard for national borders.

### Research Pays Economic Dividends

The President's Council of Economic Advisors and other economists have pointed out the high rates of return on investments in research and development. This past spring, Federal Reserve Chairman Alan Greenspan repeatedly cited an unexpected leap in technology as primarily responsible for the nation's record-breaking economic performance. In particular, a technology-based surge in productivity appears to be contributing substantially to our economic success.

This report only hints at the many non-economic benefits that result from investments in science and technology. Our military strength — built on a foundation of high technology — has enabled the United States to keep America safe from aggression, defend our allies, and foster democracies across the globe. New technologies also improve the quality of our lives. Medical research in pharmaceuticals, biotechnology, and medical devices promises us a healthier life. Environmental research offers cleaner air, water, and soil through better monitoring, prevention, and remediation technologies. Advanced monitoring and forecasting technologies — from satellites to simulation — help save lives and minimize property damage caused by hurricanes, blizzards, and other severe weather.

Agricultural research yields a cornucopia of safer, healthier, and tastier food products. Automobile research leads to cars that are safer, cleaner, more energy-efficient, and more intelligent. Aeronautical technology makes air travel safer. Energy research delivers cleaner fuels and reduces American dependence on foreign resources. Information and telecommunications technologies enable instantaneous communications across the globe.

## PCAST Investment Principles

As it advises the President, PCAST supports a steady funding stream for scientific and technological advances to help ensure prosperity and well-being for our children and grandchildren. The following principles — first adopted by PCAST in 1995 to guide the nation's investment in science and technology — remain vital to our nation's success in the 21st century.

- Science and technology are major determinants of the American economy and quality of life.
- Public support of science and technology is an investment in the future.
- Education and training are crucial to America's future.
- The Federal government should continue to support strong research institutions and infrastructure.
- The Federal investment in science and technology must support a diverse portfolio of research, including both basic and applied science.
- Stability of funding is essential.

Federal investment played an important role in the development of all the technologies described in this report — particularly in cases where the results of research were not clear at the outset. For example, when Federal funding began for the forerunners of today's Internet, no one knew what computer networking could accomplish — or how far and how quickly it would spread. The same is true of the Federal role in the development of lasers and global positioning systems. All of these investments, made because the research was fundamentally important and was needed to fulfill government agency missions, eventually changed the way businesses and governments operate and enhanced our daily lives.

The Federal government also plays a vital role in ensuring the scientific and technical literacy of the U.S. population. In our rapidly changing economy, continual learning must become a way of life. All Americans need a solid education in science, mathematics, and technology to gain the skills necessary to participate in tomorrow's workforce, to stay informed about issues, and to understand our increasingly complex world. Additional challenges for education today include a rapidly diversifying population, growing international commerce and cultural influences, an information technology revolution, and a heightened pace of change in the workplace. The Federal government, in partnerships with the states, business and industry, and academic institutions, has undertaken an ambitious agenda to foster lifelong learning — from K-12 through undergraduate and graduate school, and on to workplace training. To fulfill this task, we must employ the best educational strategies, based on solid research.

The members of PCAST, closely linked to all the stakeholders in the national and international science and technology enterprise, represent a network of advice and support that contributed greatly to this report. There is much to admire in the multiple reports issued by various stakeholders in the past few years calling for a renewed commitment to investment in science and technology. This report illustrates some of the payoffs from those investments and underscores the need for sustained and cooperative support in the 21st century to avoid the dangers and seize the opportunities.

# Chapter 1

# Information Technologies

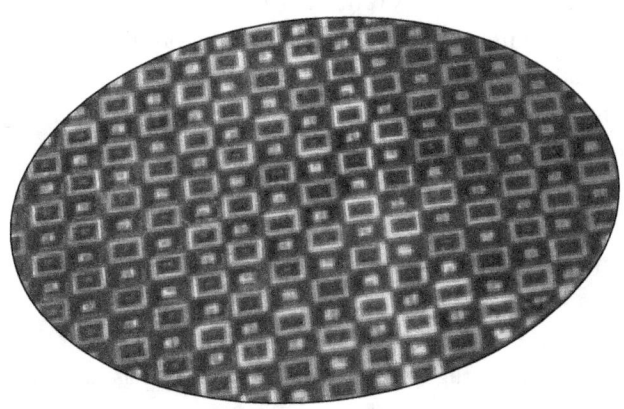

Americans are so accustomed to the presence of computers in our daily lives that we are amazed to discover that early computer experts did not foresee much demand for these specialized machines. For example, the chairman of IBM, Thomas Watson, stated in the 1940s, "There is a world market for maybe five computers." Since then, a succession of breakthroughs has created a new economic sector: Information Technologies (IT) — the ever growing variety of ways in which we are able to gather, store, analyze, share, and display information.

Estimates of the total number of computers in the world now exceed 500 million, or more than triple the total in 1991. The proportion of U.S. households with computers has jumped from 15 percent to more than 40 percent. U.S. schools now count one computer for about every six students, up from one machine for every 63 students in the mid-1980s.

The economic implications of this growth are unmistakable. Between 1995 and 1998, producers of computer and communications hardware, software, and services accounted for an average of 35 percent of the U.S. gross domestic product (GDP). During this same period, IT represented 60 percent of U.S. corporations' capital investments. On average, 20 U.S. technology and telecommunications companies are born every day. Industries that are either major producers or intensive users of IT products and services will employ half the U.S. workforce by 2006. These jobs will pay significantly higher wages than jobs in other sectors. For example, in 1996, the 7.4 million people employed in the IT sector and in IT-related jobs across the economy earned an average of about $46,000 per year, compared to an average of $28,000 for jobs across the entire private sector.

## Historical Importance of Federal Funding

Much of the innovation that spawned today's information technologies resulted directly from Federal investment in science and technology. Starting in 1969, when the Department of Defense opened its experimental nationwide computer network through the Advanced Research Projects Agency (ARPA), computer networking has especially benefited from Federal research and development funding. The National Science Foundation (NSF) extended ARPA's network to civilian academic users in 1987. These networks marked the convergence of computing and communications, one of the main drivers of information technologies in the 1990s. Networking — linking computers together to share data — has since become one of the fastest growing areas of computing. The Internet emerged from the joint effort by Federal agencies and universities to advance networking technology.

Federal agencies also contributed to the steady decrease in the size of computers and the exponential growth of computing power. The transformation from expensive, room-size computers to laptops vividly illustrates the way science and technology, through a series of chance events, sometimes lead us in unforeseen directions.

The period between 1947 and 1960 saw many separate but ultimately related breakthroughs that combined to produce a revolution in computer hardware. Electrical engineers invented the transistor and later developed the integrated circuit; materials scientists discovered semiconductors and the uses of silicon; and physicists developed the laser. The laser, once seen as an interesting but unpromising technology, has become the signal beacon that we use to transmit information, carrying light waves over optical-fiber networks that connect many parts of the globe. The laser also led to fundamental advances in lithography, the process used to etch transistors onto silicon chips to form modern microprocessors. In 1949, *Popular Mechanics* predicted: "Where a computer like the ENIAC is equipped with 18,000 vacuum tubes and weighs 30 tons, computers in the future may have only 1,000 vacuum tubes and weigh only 1.5 tons." In the days before the transistor, this may have seemed a daring prediction, but today we are on the verge of technology that will create microprocessors as small as a molecule.

### Rapid Pace of Improvements

Miniaturization causes prices to spiral downward, making the technology affordable for larger numbers of small businesses, local governments, schools, libraries, families, and individuals. During this decade alone, the cost of microprocessors has dropped sharply, as has the price of silicon-based computer memory. In 1991, the cost of the processing power required to perform a million instructions per second — or MIPS, a standard performance metric — was

## The Internet Economy

The Internet's rapid growth in popularity is unparalleled in the history of communications. Radio existed for 38 years before 50 million tuned in; TV took 13 years to reach that benchmark. The Internet crossed that popularity threshold within four years after opening to the general public in 1991, and today an estimated 80 million Americans have access at home or at work.

The Internet has quickly become a significant economic force, offering a new avenue for consumer and business-to-business transactions. The U.S. "Internet economy" grew at a compounded annual rate of 174.5 percent between 1995 and 1998, as compared with 2.8 percent for the national economy as a whole. The Internet economy generated revenues totaling an estimated $301 billion in 1998. Employing more than a million people, the Internet economy now rivals the automobile industry and other major established sectors in size. Retail sales on the Internet are also climbing rapidly. In 1997, private analysts forecast that the value of Internet retailing could reach $7 billion by 2000. In fact, the 1998 level was 50 percent greater than that estimate, causing analysts to revise their estimates upward to between $40 billion and $80 billion by 2002. Direct, business-to-business commerce on the Internet is forecast to surpass $1.3 trillion per year by 2002.

## Forecasting Severe Weather

On May 3, 1999, more than 70 tornadoes rampaged across Oklahoma and Kansas. Several days earlier, National Weather Service (NWS) forecasters had seen increasing indications in their computer modeling data that the environment in their region would support severe storms in the coming days. (NWS supercomputers continuously feed such data to field offices around the country to help them make twice-daily forecasts.) As the Oklahoma and Kansas storms developed on May 3, the NWS staff was thoroughly prepared for the intense work that lay ahead to analyze the storms and predict their likely paths, enabling real-time forecasts. As a result, some areas were warned of the impending fury up to 60 minutes before the arrival of severe weather.

More than 1,000 tornadoes occur in the United States every year. Since the early 1990s, technological progress in detecting, monitoring, and modeling such storms has doubled average lead times for tornado warnings, from a nationwide average of less than four minutes to more than 11 minutes. Using the sophisticated Advanced Weather Interactive Processing System, which is now installed at NWS forecast offices throughout the United States, forecasters can process, display, and integrate huge amounts of storm data much more efficiently, adding to the speed and accuracy of warnings to emergency managers, the media, and the public.

Ultimately, with continuing upgrades of technology and progress in science, the NWS expects to deliver reliable warnings at least three hours before the onset of severe weather.

*Advance warning of this May 3, 1999 tornado, made possible by NWS technology, enabled officials to order the safe evacuation of millions of residents.*

$230. By 1997, the cost of one MIPS' worth of computing power was $3.42. As a result, today's desktop and laptop computers pack the processing power equivalent to what would have been a supercomputer only a few chip-generations ago. Advances also continue at the high end of computing. State-of-the-art supercomputers can perform trillions of operations per second. Federal research in high-performance computing envisions machines within the next ten years that will be thousands of times more powerful, performing a quadrillion operations per second.

Vast increases in computing power have led scientists and engineers to tackle problems once considered beyond the limits of human study — such challenges as long-range weather forecasting, modeling thermonuclear explosions, simulating variations in aeronautical design, and designing new drugs. These efforts depend on improvements in computer software as well as hardware. Mathematics holds the key to the efficient trade-offs in processing speeds, message passing, and use of memory that make up programming code. Recent advances in this area of science have generated a proliferation of software for professional and everyday uses. In terms of economic impact, the software and computer services sector has more than doubled in size since 1990, growing to a $152 billion business by 1998 and roughly equaling the size of the computer hardware sector.

The industrialized world has integrated information technologies throughout almost all economic sectors and social institutions. We use IT when we listen to a weather forecast,

## Libraries With Instant Access

A $30 million cooperative initiative of several Federal agencies will soon provide a national library of text, images, sound recordings, and other materials to every schoolchild (and all other Americans) with access to the Internet. The library will include countless numbers of books that are in the public domain, such as the complete works of Shakespeare, Mark Twain, the Greek philosophers, and the Federalist Papers. In addition, children will have virtual access to such things as the Apollo 11 command module, the Gettysburg battlefield, and Rose Kennedy's personal tour of the John F. Kennedy birthplace.

Students and their teachers will also be able to find a digital library for math and science education as part of the new initiative. The collection will include high-quality resources and provide hands-on, interactive content that makes math and science come alive and enables students to "learn by doing."

Technology can help make America's treasures available to all citizens. It will still be a thrill to visit Ellis Island in person, but those who can't make the trip will still have online access to its immigration records. Poetry lovers will still want to own a personal copy of Dante's *Inferno*, but every student who needs to read it will be able to download it instantly — and never have to pay an overdue fine!

*Digital libraries will provide all Americans with unprecedented access to information.*

enjoy music on a CD, or watch a movie on DVD. The free exchange of information made possible by the Internet has had a democratizing influence in other parts of the world, and small but innovative entrepreneurs now have opportunities to market their ideas on a worldwide scale that would have been impossible to imagine 20 years ago.

## Information Technologies of Tomorrow

The evolution of information technologies has surpassed expectations for more than a half-century, and the next 50 years hold even greater promise. As researchers improve the intelligence capabilities of computers — particularly the ability to imitate the human process of reasoning — they will become an integral support in key decision processes. Decisions that include large numbers of factors involving natural and human processes, such as those on the military battlefield, will benefit from this capability. Soldiers will carry wireless devices with sensors that will "read" their surroundings — physical terrain, air quality, and compass direction — as well as the soldier's own vital signs. A tiny transmitter will continuously relay the data back to a command post for analysis by larger computers as part of an ongoing, real-time decision-making process. As a result, commanders will have many more options available at an earlier stage, improving the chances for military success.

Closer to home, we need only consider that there are an estimated 20 billion microprocessors embedded in products around the world other than computers: our cars, watches, air conditioners, microwave ovens, and VCRs. Soon we will live in "smart houses," where we will plug dishwashers, clocks, stereos, and other household appliances into the Internet along with the electric power grid. At that point, instead of e-mailing home to ask the family to start dinner, we'll be able to e-mail our kitchen appliances directly and order them to move the lasagna from the refrigerator to the oven and cook it at 375° for 45 minutes. The marvels of the Information Age are limited only by the imagination of future engineers and consumers.

## Chapter 2
# Global Positioning System

What would it be worth if you could pinpoint your location anywhere on Earth, whenever you might need to know it, to an accuracy within a few hundred feet, using only an inexpensive handheld device? Would you answer this question differently if you were an airline pilot concerned that you had drifted off course? What if you were general manager of a trucking company trying to save time and money as you schedule the next two days' pickups and deliveries for a fleet of 200 long-haul trucks? What if you were a weekend sailor sending a distress radio call because you're in danger?

During the past decade, real people like those described — and thousands of others in a wide range of scenarios — have relied increasingly on the Global Positioning System (GPS) to calculate precise location. This ability, undreamed of before the age of satellites, has already saved countless lives and dollars for those who use it.

GPS consists of a constellation of 24 satellites that orbit the Earth every 12 hours, each emitting radio signals coded with data about its position and the time — accurate to within a billionth of a second. The satellites are deployed so that every point on Earth can always receive signals from at least four satellites. Receivers on Earth interpret these signals to pinpoint their own positions, anywhere on the globe, at any time of day or night, in any kind of weather.

### A Breakthrough Based on Basic Research
The GPS we use today would not have been possible without basic scientific research in atomic and molecular physics or advances in satellite, launch vehicle, and telecommunications technology. GPS technology actually grew out of pure physics research, starting in the 1930s, by scientists studying the nature of the universe and how to measure time exactly. By the 1950s, this research had developed extremely accurate atomic clocks, which would later be crucial in developing the concept of GPS. The advent of space satellites, with the Soviet Union's launch of Sputnik I in 1957, allowed scientists and engineers to envision a system of navigation that would rely on satellite signals keyed to precise timekeeping. By 1973, the Department of Defense had approved the navigational concept that became GPS. Rockwell International began building the Navstar satellites that make up the GPS constellation, each the size of a large automobile and weighing slightly less than a ton. In 1978, the first GPS satellite became operational; by 1993, the full 24-satellite system was in use. Technological advances in solid state electronics, microchips, and microwaves also contributed to the commercial success of GPS. In 1983, the first GPS receivers cost over $150,000 and weighed more than 100 pounds. Today, a handheld GPS receiver weighing less than a pound can be purchased for less than $100.

## Mapping Remote Regions

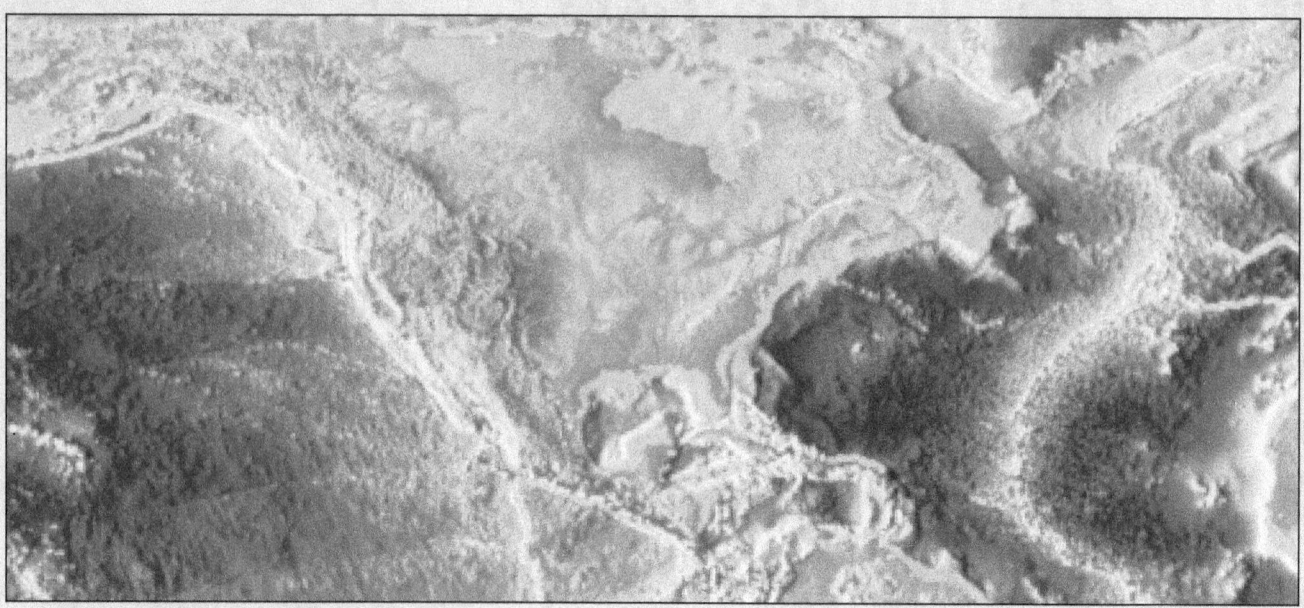

*GPS technology has become instrumental in efforts to map remote regions. The GPS network enables cartographers to create extremely accurate maps of previously uncharted territory.*

GPS has brought enormous benefits to people who live, work, and play in uncharted terrain. For example, there was no reliable map of China Camp, a 1,600-acre site in Marin County, CA, which draws more than 2,000 hikers, campers, and mountain bikers every summer. Firefighters sometimes needed an hour or more to locate a wildfire — and even more time to find a lost or injured hiker. Now firefighters ride the trails of the park on a mountain bike, using GPS data to follow a highly accurate computerized map of China Camp that can be used by visitors as well as emergency personnel. These maps have decreased the department's average emergency response time to 20 minutes.

Across the globe, in sub-Saharan Africa, malaria takes the lives of more than 1.5 million infants and children under the age of five each year. In 1995, the U.S. Centers for Disease Control and Prevention (CDC) initiated a five-year study of the disease, using GPS equipment to map the location of 450 households, three rivers, major roads (permanent and seasonal), mosquito breeding sites, medicine stores, local health clinics, a lakeshore, and other relevant features. By providing more accurate maps of locales — and even establishing clear correlations between specific dwellings and frequency of illness — the project has quickly demonstrated the practical value of GPS for researching tropical diseases in remote locations.

Most Americans first became aware of GPS when the U.S. military used it successfully in the 1991 Gulf War to target "smart" bombs and cruise missiles. In the deserts of Kuwait and Iraq, GPS gave U.S. forces a precise and reliable sense of where they were in unfamiliar territory. GPS again made headlines in 1999, when a U.S. F-16 went down in the northwestern part of Serbia during Operation Allied Force in the liberation of Kosovo. The American pilot was rescued by NATO forces and taken to safety within a matter of two hours, thanks to GPS technology.

## A Robust Economic Force

GPS is now a dual-use technology with civilian uses that rival its continuing military role. President Clinton announced in 1996 that the U.S. government would continue providing GPS signals to the world free of direct user fees, as a public good. GPS has since developed into a multi-billion-dollar international industry, creating thousands of new jobs while saving lives and bringing many other benefits. The number of companies identifying themselves as providers of some sort of GPS-related goods or services has grown from 109 firms in 1992 to 301 firms in 1997. Even though relatively few of these firms compete to provide the core GPS technology, a large number of firms provide GPS-enhanced products and value-added services. The technology has clearly carved out a crucial role for itself in the global information infrastructure. The precise GPS timing signals that help synchronize global information networks of fiber optics, coaxial cable, copper wire, radio, and communication satellites have become essential to daily commerce.

Like many of the other technologies covered in this report, global positioning — in both its history and its current uses – draws on the breakthroughs in the Information Technology field; we highlight it here mainly because of its amazing economic promise. The global GPS market, currently estimated at more than $2 billion per year, is projected to expand to $30 billion annually before 2030. GPS receivers and transmitters may soon be smaller than credit cards — and cheap enough for use in almost any vehicle, cell phone, or pocket, for that matter. With every square yard on Earth measured and labeled with an address, and with computerized databases available that give latitude and longitude as well as addresses, it's conceivable that no one will ever need to ask directions again.

## Saving Lives

Mitch Buffim of Buffalo, New York, knows firsthand what GPS can do. Buffim is one of 500 volunteers in Erie County who are testing a new GPS-based automatic collision notification (ACN) system. After working a late shift one night last April, Buffim fell asleep at the wheel while driving home on a rural road. His car ran off the road, rolled on its side, and slid down a steep embankment. With no witnesses and his car invisible from the roadway, he could have waited hours for help. However, almost before the car stopped moving, Buffim heard the emergency dispatcher's voice in the car. "This is Erie County Dispatch. We have your location. Are there any injuries? How many occupants are in your car?" Thanks to the ACN system in Buffim's vehicle, the dispatcher knew the exact location of the vehicle, its speed, and the force of the impact. Help was on the way almost instantly. Single-vehicle rural crashes like Buffim's account for one-third of all fatal crashes nationwide. ACN is connecting these crash victims with emergency care well within the "golden hour" that often means the difference between life and death. If help can arrive ten minutes sooner during this first hour, 9,000 lives may be saved across the nation each year.

*GPS technology plays an important and growing role in the provision of emergency services.*

## Improving Transportation Efficiency

Public and private organizations rely on GPS and other technologies to improve transportation safety and efficiency. The cascade of benefits includes millions of dollars in savings throughout the economy, enhanced customer satisfaction, and improved air quality. For example, each of Denver's 800 buses is equipped with a GPS-based automatic vehicle location system that reports the location of the bus every two minutes. Dispatchers have improved their ability to keep buses running on time by viewing bus locations on computer screens that are fully integrated with digital city maps. The system is credited with increasing use of the bus system, relieving traffic congestion, and reducing smog.

GPS tracking technology at the American President Line's Global Gateway South at the Port of Los Angeles automatically matches a cargo container's identification number with its location in the yard. Back at the terminal, a computer stores the GPS location and content data for each container. Using on-board navigation, drivers can now negotiate the 6,000-space holding tank and drive straight to the proper container the first time, eliminating costly mistakes and saving time and money. The system increases the overall efficiency of the cargo storage space, an important benefit for port facilities with no room for expansion.

## Chapter 3
# Biomedical Technologies

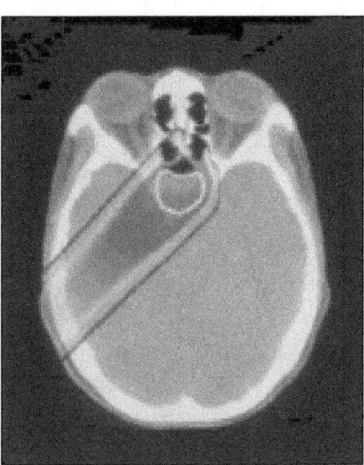

The United States has seen amazing changes in biomedical technologies over the past 100 years. We have come from the family doctor's signature black bag in the first half of the century to the powerful scanning equipment of the modern medical center; from surgical saws to the lasers, endoscopes, and angioplasty of today's operating rooms; from tens of thousands dying in influenza epidemics to hundreds of thousands of seniors receiving their annual flu shots; and from an average life expectancy of about 49 years to our present expectancy of 75 years.

Medicine saves lives and relieves suffering. It embodies for many of us the greatest achievements of science and technology. The rapid progress in medicine has come from life sciences — such as biology and genetics — but also from physics, math, and many other fields of science and engineering.

### Contributions from Physical Sciences

For example, over the past 25 years physicists have developed revolutionary imaging technologies that have allowed us to see deeper and deeper into the materials and processes of life itself. Doctors are now using non-invasive means of looking into the human body to diagnose a wide variety of diseases — including cancer, multiple sclerosis, Alzheimer's disease, stroke, heart failure, and vascular disease. CAT (Computer-Assisted Tomography) scans combine X-rays with computer technology to create cross-sectional images of the patient's body, which are then assembled into a three-dimensional picture that displays organs, bones, and tissues in great detail. Magnetic Resonance Imaging (MRI) scanners use magnets and radio waves instead of X-rays to generate images that provide an even better view of soft tissues, such as the brain or spinal cord. Ultrasound images, produced by very-high-frequency sound waves, can help doctors visualize a developing fetus, detect tumors and organ abnormalities, and identify women at risk of developing osteoporosis. Imaging technologies have also greatly helped in early detection of breast cancer, which claims the lives of nearly 42,000 American women each year. The deeper and smaller we see, the more we understand how life processes work on their most fundamental level.

Mathematics and computer science have greatly contributed to biomedicine through information technology. Much of today's imaging technology relies on microprocessors and software. Computers are also making it easier for researchers to collect, analyze, and share data in research and in telemedicine, and to model biological systems to project likely outcomes more accurately. It would be impossible for scientists to sequence the entire human genome without the information processing power of supercomputers. And information technologies have provided essential tools to collect and analyze data for epidemiological research that helps us understand the distribution of disease and to develop clinical and public health interventions.

Another development from the physical sciences, the laser, has made the scalpel unnecessary in many kinds of surgery. Laser surgery reduces pain and trauma for the patient, speeds healing — thereby shortening costly hospital stays — and improves the accuracy of certain surgical procedures. Most notably, eye surgery has been revolutionized by this new technology. Precision lasers have been used to halt, and in some cases reverse, diabetic retinopathy, a dangerous complication of diabetes and the leading cause of new cases of blindness in adults. Lasers can also be used to repair small tears in the retina, preventing retinal detachment, and also to provide follow-up treatment to patients after cataract surgery. Most recently, ophthalmologists have begun to use lasers to correct nearsightedness, in a procedure called LASIK (laser in situ keratomileusis). Not only is laser eye surgery effective, but it is fast and relatively painless.

## Contributions from Life Sciences

Of course, the biomedical revolution also sprang from fundamental advances in our knowledge of the life sciences, particularly knowledge of genetics. Between 1665, when Robert Hooke first observed cells, and the middle of this century, researchers learned that heredity is controlled by genes, that genes are located on chromosomes, and that genes are made from deoxyribonucleic acid (DNA). In 1953, Watson and Crick discovered that the structure of DNA, which is common to all life on Earth, is a double helix. That breakthrough swiftly cascaded into new techniques that allow researchers and clinicians to control biological processes in very precise ways.

Today industrial-scale production of insulin for diabetics is possible because scientists learned how to cut and paste the human insulin into bacteria that can produce large quantities of the substance inexpensively. Gene transfer

### Hello... Is the Doctor In?

Millions of Americans received their first introduction to telemedicine in the summer of 1999 by following the news story of Dr. Jerri Nielsen, 47. Dr. Nielsen, who was serving at the U.S. Amundsen-Scott South Pole research station, discovered a lump in her breast during a routine self-examination. She conducted telephone consultations with doctors via satellite. On their advice, medical supplies were air-dropped, with which Dr. Nielsen treated herself for several months until warmer weather permitted her to be airlifted out safely.

Less dramatic telemedicine occurs daily. Some doctors regularly e-mail medical images such as CAT scans to colleagues for review. In remote rural areas, telemedicine can mean the difference between life and death. For example, a specialist at a North Carolina University Hospital was able to diagnose a patient's hairline spinal fracture at a distance, using telemedicine video imaging. The patient avoided paralysis because treatment was done on-site without physically transporting the patient to the specialist, who was located a great distance away.

As the practice of telemedicine spreads, doctors may be speaking literally when they say, "Call me in the morning and let me know if you feel better."

*Dr. Jerri Nielsen was a part of the National Science Foundation-funded research mission in Antarctica.*

techniques are also used to produce antibodies that can attack cancerous tumors directly or deliver lethal doses of drugs to tumors without damaging surrounding tissue. Many of today's vaccines — which save $6 to $16 in medical costs for every dollar spent on production — come from genetic engineering.

Knowledge of genetics will be further extended by the Human Genome Project, an ambitious international effort to determine the complete human DNA sequence, funded by the National Institutes of Health, the Department of Energy, and the United Kingdom's Wellcome Trust. A map of the human genome published in October 1998 contains over 30,000 genes, almost twice as many genes as the map published in 1996. The work of the Human Genome Project has led to development of tests that doctors are already using for screening and diagnosing disease.

The HGP includes an important new research component that focuses on the ethical, legal, and social implications (ELSI) of genetic research. This program will help ensure that developments in genome science and technology take account of values such as privacy and affordable health care. The ELSI program also will serve as a model for other technological initiatives that raise concerns about established cultural norms even as they offer tremendous advantages.

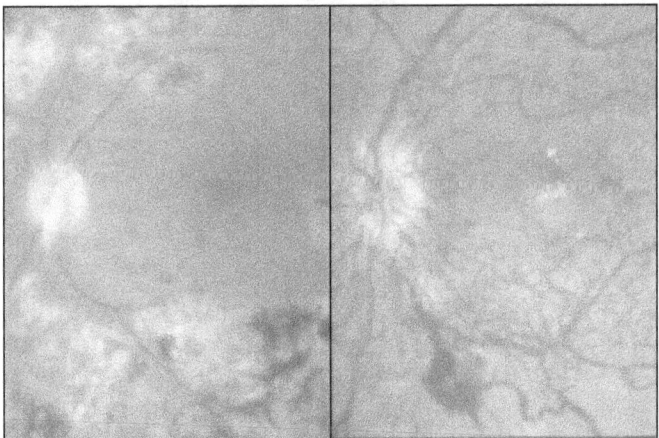

*Improving the health of all Americans requires a broad spectrum of basic research across all the scientific disciplines, often drawing upon tools developed in the physical sciences. Here a laser is used to treat eye disease, before (left) and after (right).*

## A Powerful New Prevention Tool

The vaccine against Hemophilus influenza type b (Hib) meningitis provides the means to completely eliminate this disease from the United States within the next few years. This turnaround is largely a result of basic scientific research in molecular biology. For years this disease struck 15,000 to 20,000 U.S. children each year — almost as many as polio at its peak. It killed 10 percent and left one-third deaf and another one-third mentally retarded, making it this country's leading cause of acquired mental retardation. Fortunately, two NIH scientists made a discovery about how to make infants' bodies fight the disease, a discovery that led to the development of a safe and effective vaccine. The vaccine, routinely administered to babies only two months old, is saving more than $350 million per year in avoided infections, and the incidence of Hib has declined by 95 percent since 1988. With greater use of the vaccine across the country, we have the hope of completely eliminating Hib meningitis.

### Healthy People, Healthy Economy

Just as biomedical technologies have made enormous contributions to Americans' health and well-being in this century, they have also helped the economy. The health care industry generates roughly $1 trillion in economic activity, high-wage jobs, and trade. Another measure of medicine's economic importance is the amount of money it saves: for example, improved treatment of acute lymphocytic leukemia has saved the nation more than $1 billion in restored lifetime earnings, and lithium treatment for manic-depressive illness has saved about $7 billion per year since its introduction in 1970. And the list of cost savings continues to grow.

The challenges facing biomedical sciences in the 21st century are daunting. Emerging infectious diseases such as AIDS are a major threat across the globe. Antibiotic-resistant strains of infectious agents threaten progress already made against diseases such as tuberculosis. In this country, coupling prolonged good health with prolonged life span remains an unfinished task. Today's killers and disablers more often arise as a consequence of things we do to ourselves (unhealthy behaviors such as smoking,

## Healthy Hearts—Right From the Start

Over the past two decades, medical science has managed to reduce deaths from stroke by 59 percent and deaths from heart attack by 53 percent. One major reason for this success has been the development of drugs that combat hypertension. A concentrated research effort that combined the efforts of the Federal government, pharmaceutical companies, voluntary health agencies, and private foundations contributed to this feat. Although these decreases in deaths are encouraging, we still don't know enough about how hypertension works. Preventing this condition is still an elusive goal.

In addition to modern drug therapy for heart disease patients, medical scientists consistently advise careful eating habits, since diet can contribute to the risk of cardiovascular disease. The long-established eating habits of adults can be extremely resistant to change. But it may be possible to teach younger Americans to eat more nutritious foods. A study supported by the National Heart, Lung, and Blood Institute at the National Institutes of Health suggests that an intensive school and family-based intervention program can have lasting effects.

More than 5,000 grade-school students from nearly 100 ethnically and racially diverse elementary schools in California, Louisiana, Minnesota, and Texas participated in the original CATCH (Child and Adolescent Trial for Cardiovascular Health) Study between 1991 and 1994. The children learned to read labels; to select "Go," "Slow," and "Whoa!" foods; and to prepare healthy snacks. They ate heart-healthy school lunches, participated in more moderate to vigorous activities in PE classes, and engaged their families in entertaining activities and games promoting healthy eating and exercise behaviors.

In a follow-up study, researchers found that the students who received the health promotion intervention in grades three through five maintained a diet significantly lower in total fat and saturated fat and continued to pursue more vigorous physical activity levels than did students in the control groups. These results suggest that schools can be an important place to help young people establish habits that may help prevent the early onset of cardiovascular disease — the leading cause of death among Americans.

*Nutrition plays a pivotal role in reducing the risk of cardiovascular disease.*

drug and alcohol abuse, poor diets, failure to adhere to drug regimens, or inadequate physical exercise) or others (violence and injuries). Growing evidence that changing behavior reduces the risk of disease suggests that our efforts to improve human health must address the complex interplay between body and mind.

A broad portfolio of balanced research investments is the key to advancing biomedicine. The physical, mathematical, behavioral, and other sciences must continue to advance in tandem with the life sciences if we are to continue to make progress against disease.

## Chapter 4

# Food Technologies

Walk into almost any grocery store in the United States, and you will be overwhelmed by the sheer variety of foods. All year round, fresh fruits and vegetables, dairy goods, meat and poultry, baked products, canned goods — and, of course, snack foods — line the aisles. Not only are our supermarket shelves well stocked, but we also export enormous amounts of agricultural products to the rest of the world. Sometimes we take this abundance for granted, but we can thank science and technology for these blessings.

As the quantity of food available has increased, overall food prices have steadily decreased, so that food in the United States is also more affordable to the consumer. According to the U.S. Department of Agriculture, the average U.S. family spent 10.7 percent of its income on food in 1997, compared to 11.6 percent a decade earlier and 22 percent 50 years ago.

## Progress Through Research

The United States has increased its agricultural output largely through growth in productivity, which rests heavily on our long history of Federal investment in agricultural research, development, and infrastructure. Economists have found that the annual rate of return for publicly funded agricultural research is about 35 percent — for every dollar spent, society gains $1.35 in benefits.

Research in the agricultural sciences is continuous and cumulative. Our modern supermarket cornucopia did not spring forth overnight or as a result of research in any one field of science. Rather, over many decades of research, our farmers and scientists have adapted knowledge from many scientific disciplines to help them grow and deliver more nutritious and satisfying food to our citizens with less harm to the environment.

Remarkable advances in genetics, for example, have steadily pushed agriculture forward. At the beginning of the 20th century, scientists were rediscovering work done some 30 years earlier by the Austrian monk Gregor Mendel, who conducted breakthrough scientific experiments proving that plant traits are largely inherited. Mendel's research allowed scientists early in this century to develop selective breeding techniques, which identify agronomically desirable genetic traits and integrate those traits into crops and livestock to improve them.

## The Promise of Genomics

Further advances in genetics, coupled with powerful computer technology, led to a new field of study called genomics — the study of the entire DNA complement of an organism. Using traditional breeding practices, scientists found the means to move desirable genes between sexually compatible species. Today, scientists can use genetic engineering to move genes between unrelated species — and can modify them to function in specific tissues at specific times.

In 1999 alone, U.S. farmers planted roughly 25 percent of the nation's corn crop using genetically engineered varieties that reduce our dependence on toxic chemical pesticides. Half of all soybeans planted in the United States in 1999 will be seeds that have been genetically modified to resist herbicides. New varieties of fruits and vegetables are being grown that will ripen more reliably and resist virus infections better.

## Improving Farm Productivity

Farmers increasingly rely on precision farming — tools and techniques designed to work the land by the square meter instead of the square mile — to improve productivity. Using GPS and other technologies, farmers can now achieve an extraordinary degree of accuracy in a range of operations including field mapping, soil sampling, fertilizer and pesticide application, and crop-yield monitoring.

Thayne Wiser's 2,000-acre farm lies in the "rain shadow" of Washington's Cascade Mountains, and receives only six inches of rain a year, mostly in winter. Irrigation is critical to the growth of his crops, but managing it is a complex process. The soil on Wiser's farm is sandy and prone to erosion and leaching, and it varies in its ability to hold water and nutrients. To complicate matters, the water pressure at the hundreds of sprinkler heads that make up Wiser's irrigation system varies with elevation. Balancing soil moisture levels with the right amount of fertilizer and pesticide had been more of an art than a science, but GPS helped Wiser develop a precision irrigation system that saves water, reduces runoff of pesticides and fertilizers from fields, and increases crop yield.

*Scientists provide farmers with information on new pest management practices that meet agricultural production, human health, and environmental goals.*

Genomics research will be particularly important to the emerging U.S. aquaculture industry. Declining natural fishery harvests and rapidly growing populations mean that aquaculture production will need to increase some 300 percent worldwide by 2025 to meet projected seafood demand. There is great potential for rapid gains in growth rate, production efficiency, and health status of farm-raised fish through targeted genomic research.

In some parts of the world, genetically engineered foods have generated environmental and human health concerns. In the United States, our food safety regulatory agencies have a strong track record of utilizing sound science in their regulatory actions, and that engenders public trust. Citizen participation in the policymaking process and ongoing research and outreach programs to address emerging concerns can help ensure that these new food technologies serve the best interests of consumers.

## A Safer Food Supply

There is almost universal agreement that the food we eat today is safer than that of any previous time. When we bite into a hamburger or munch on fresh fruits and vegetables, we are not generally putting our health at risk. This is largely a tribute to our nation's food production system, which reduces the possibility of exposing us to harmful pathogens or chemicals, all the way from farm to table. Although our food supply is one of the world's safest, we still need to do better. Every year far too many people, especially the very young and the elderly, become ill or die as a result of contaminated food.

## DNA: Detective for Food-Borne Pathogens

After DNA "fingerprinting" was successfully used in 1995 to stop an outbreak of *E. coli* illness, the Centers for Disease Control and Prevention (CDC) established a national network of public health laboratories to track foodborne bacteria. PulseNet laboratories in 22 states and two major cities can quickly identify and compare specific DNA patterns found in bacteria isolated from sick persons or contaminated foods by using the same technology that creates DNA fingerprints of human criminals. The technique creates a "barcode" pattern, unique to each type of bacteria, that can be quickly compared to the barcodes of bacteria in the CDC's centralized electronic DNA database. This tool helps us understand how foodborne illnesses might be spreading from a common source and how to stop them.

PulseNet now plays a vital role in surveillance and investigation of foodborne illnesses that were previously difficult to detect. Scientists can spot an outbreak even if its victims are far apart geographically. With new fingerprinting tools, electronic technologies, and Federal coordination through the CDC and states, outbreaks and their causes can be figured out in hours rather than days, and control measures can be instituted more quickly to prevent loss of life and illness.

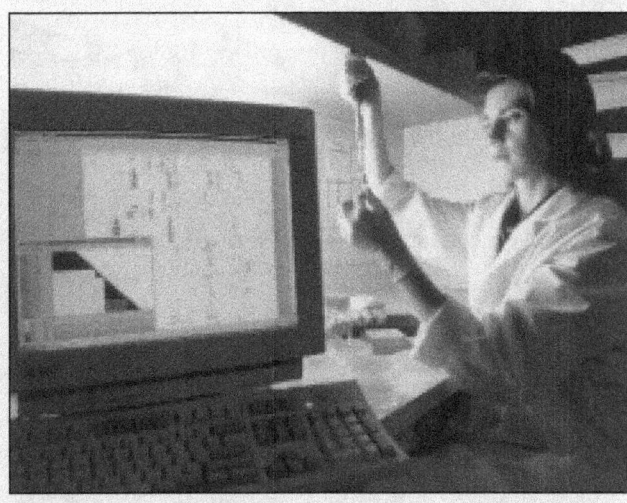

*A strong research program in food safety ensures public health and reduces the prevalence of food-borne diseases.*

Recent advances have helped scientists better understand the role of pathogens in food contamination, creating new opportunities to improve the safety of our food supply. There is a new emphasis on preventing contamination throughout the entire production process, rather than simply inspecting food at the end of the cycle. Tests for pathogenic bacteria that once took days to complete now take minutes. Researchers are developing biosensors that will be used to indicate whether a product on the shelf in the store is suitable for consumption. With research efforts such as these tied to strong regulatory programs, we can expect to see a significant reduction in cases of foodborne illnesses over the next decade.

When contaminated foods do cause illness, new tools help us control those threats. For example, thanks to breakthroughs in computer networking technology and genomics, we are now able to track the DNA fingerprint of specific pathogens — the telltale genetic code that proves the identity of a contaminating organism. With that information in hand, we can quickly determine whether its appearance is related to other outbreaks and even whether separate incidents can be traced to the same source in the food supply and distribution chain.

### Food Technologies of the Future

New technologies may soon be used to engineer functional foods to be eaten for specific needs. Vitamin A-rich produce will be grown in developing countries, where a deficiency of this nutrient causes a half-million children each year to become permanently blind. Future "nutraceuticals" will likely include staple foods, such as potatoes and bananas, genetically engineered to contain vaccines that will stimulate human antibodies against disease.

Advances in food technologies promise to continue today's trends toward healthier food, available in a wider variety of convenient forms, delivered more safely to our table, and produced with less impact on the environment. The farmer will remain at the center of our food production system, but increasingly science and technology will give farmers — and consumers — new confidence in a bountiful harvest.

## Biodiversity as the Foundation of Agriculture

Whoever first said "variety is the spice of life" may have been wiser than we thought. Judging from what scientists have established over the past 50 years, variety in and among plant and animal species is what allows life itself to survive and thrive. Biological diversity, or biodiversity — all the species on Earth, all the varieties within each species, and all the ecosystems that sustain them — is now recognized as a critical factor in the natural processes at work in agriculture.

When we began domesticating plants and animals, we wrought perhaps the most far-reaching single change in the ecology of the Earth to that time. By patiently perfecting food crops such as wheat, rice, and corn and making them the dominant crops around the world, we also enabled humans to emerge as the dominant animal species. Since then, we have steadily improved our standard of living despite ever greater population densities. Agriculture has permitted this progress, and agriculture relies on biodiversity for its continued success.

Biodiversity allows higher yields, pest resistance, and improved quality of crops. It fosters the development of crop varieties that adapt to different soils,

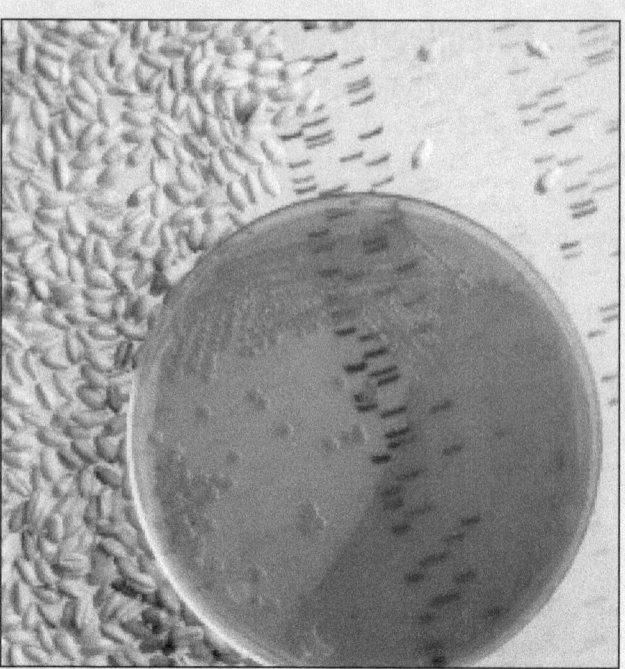

*Wheat seeds treated with bacteria like those colonized in this petri dish are nearly immune to wheat take-all, a root-destroying fungal disease. The sequencing gel in the background bears the genetic code for bacterial enzymes that synthesize natural antibiotics.*

climate regions, and environmental threats such as insects and disease. Of the annual increases in crop productivity achieved through agricultural research, about half are attributable to extractions from wild species in biodiversity's "genetic library."

Biotechnology now offers us the ability to tap into the genetic diversity of all species — not just close relatives — and apply desirable traits to completely different species. For this reason, preserving natural biodiversity is more crucial than ever before. We cannot predict which individual strain — or even which particular plant species — might, at some point in the future, offer a genetic weapon against a pathogen or pest and thereby save millions of acres of food crops from ruin. And although scientists are now able to move genes from species to species, only nature can create them. Only by maintaining the widest possible diversity among all species and their related ecosystems can we hope to ensure that we will have the resources to develop new crop varieties when needed to respond to environmental challenges.

*New molecular biological techniques are expanding the possibilities for developing healthier, disease resistant plants.*

# Chapter 5

# Environmental Technologies

During the past 30 years, environmental science has dramatically altered our perception of the relationships between human activity and the natural environment. Back in the 1950s, it would have seemed absurd to suggest that humans could in any way alter the global cycles that sustain life on our planet. In particular, the oceans and the atmosphere — our two global commons — seemed to be inexhaustible resources whose very vastness made them invulnerable to influence by humans.

Today we know this view was wrong. Global population has more than tripled over the last century, and human expectations have risen steadily. Consumption of natural resources by the industrialized world has risen to heights undreamed of even a few decades ago. In just a global instant, the world has ceased being "wild." It is estimated that humans now consume or divert nearly half of the net plant productivity of the land, use more than half of the available fresh water, and significantly modify the composition of the atmosphere.

Environmental science has given us a better understanding of the complexity of the Earth's environment and its sensitivity to stresses caused by a growing human population. It has also revealed much about human dependence on the healthy functioning of those ecosystems for food, timber, clean water, medicine, and recreation. But we are still learning painful lessons; for example, the devastating floods that China suffered in 1998 were partly a result of extensive deforestation in critical watersheds.

## The Goal of Sustainability

New knowledge has led to a new emphasis on developing sustainable uses of natural resources. The challenge is to enable development — including economic growth — without harming the natural environment. Sustainability requires consideration of complex interactions: maintaining biological diversity, safe water resources, and air quality; protecting the population from toxic substances and natural hazards; reversing stratospheric ozone depletion; and understanding, mitigating, and adapting to climate change.

Developing sustainable practices requires a comprehensive scientific understanding of the environment and the development of innovative and creative new technologies to help solve those problems. There is no better example of this process than the story of chlorofluorocarbons (CFCs). As early as the 1960s, scientists were beginning to understand that these industrial chemicals — widely used because of their many desirable chemical properties — pose a threat to the thin layer of stratospheric ozone that protects life on Earth from dangerous levels of cancer-causing ultraviolet radiation from the sun. As this threat was more fully documented, scientists and engineers from the government and the private sector helped solve the problem by developing safer, less ozone-destructive substitutes for CFCs. Thanks to those efforts, the world was able to sharply reduce CFC use years earlier than originally thought possible. Today, those efforts are paying off: atmospheric measurements show that levels of CFCs in the stratosphere are already leveling off.

## Recent Progress in Cleaning Up

It is difficult today to imagine the levels of air pollution that were commonly accepted in the 1940s and 1950s as the inevitable price of industrial progress. After a particularly acute episode of air pollution in London in 1952 killed some 4,000 people, scientists went to work to understand the sources of air pollution. Some of the sources, such as coal-fired boilers, were readily identified. But "smog" was more difficult. Atmospheric scientists eventually determined that sunlight shining on exhaust from tailpipes and smokestacks causes smog. Knowledge of the cause led environmental engineers to solutions such as catalytic converters for automobiles and scrubbers for industrial smokestacks. For example, today's cars get twice the average gas mileage of cars built in 1970, and they burn their gasoline 90 percent more cleanly. Since 1970, air pollution has declined by 31 percent, while U.S. population increased by 31 percent, GDP increased by 114 percent, and vehicle miles traveled increased by 127 percent.

Just as science and technology led the way in improving air quality, they have also given us new understanding, and new tools, in the effort to clean up our water. The

### East Side, West Side, All Around the Watershed

New York City has had a long tradition of supplying clean municipal water. New York's water, which originates in the Catskill Mountains, was once bottled and sold throughout the Northeast. In recent years, sewage and agricultural runoff have overwhelmed the Catskills' natural ecological purification system, and water quality dropped below EPA standards. This prompted New York City's administration to investigate the cost of replacing the natural system with an artificial filtration plant. The estimated price tag for this installation was $6 to 8 billion in capital costs, plus annual operating costs of $300 million — a high price for a commodity that was once virtually free.

Further investigation showed that the cost of restoring the integrity of the watershed's natural purification processes would be a small fraction of the cost of a filtration plant — about $1 billion. The city chose the less costly alternative, raising an environmental bond issue in 1997. It is now using these funds to purchase and halt development on land in the watershed, to compensate landowners for restrictions on private development, and to subsidize the improvement of septic systems.

In this case, a financial mechanism has helped to recapture some of the economic and public health benefits of a natural capital asset, the Catskills watershed. The full economic and ecological value is much greater, however, since conserving the Catskills ecosystem for water purification will also protect its other benefits, including tourism and recreation, flood control, and wildlife and fisheries. Such financial mechanisms can be applied in other geographic locations and other ecosystems to benefit municipalities and habitats throughout the nation.

*Advances in environmental science and technology hold tremendous promise for creation of a sustainable future, a future where environmental health, economic prosperity, and quality of life are mutually reinforcing.*

## New Markets for Energy Technologies

Currently, fossil fuels provide more than 75 percent of the world energy supply. However, researchers expect the next century to bring strong new growth in development of renewable energy technologies — such as wind power, photovoltaic cells, and biomass — that are friendlier to the environment. Over the longer term, these "renewables" will be economically competitive with fossil-fuel technologies.

Developing countries around the world are expected to play a particularly prominent role in the rise of renewable energy technologies. In fact, over the next two decades, more than half of global energy growth will be in developing and reforming economies. Between now and 2050, investments in developing countries in new energy technologies are projected to reach a level between $15 trillion and $25 trillion. Additional investments in energy efficiency are expected to be on a similar scale as these countries create their buildings, industry, and transport infrastructures.

This dynamic new global market for energy technologies will likely stimulate new, perhaps even revolutionary, energy technologies that will allow us to continue improving the quality of human life while reducing our dependence on fossil fuels and their associated environmental dangers. This market also represents a remarkable opportunity for American businesses, if they are ready with the technologies that emerging economies demand. Federal funding of research to fill the gaps in private-sector investment can achieve significant benefits for the United States.

---

United States has 3.5 million miles of rivers, 41 million acres of lakes, 277 million acres of wetlands, and 34,400 square miles of estuaries. During the past 25 years, we have seen substantial improvements in water quality for many types of pollutants in the nation's aquatic ecosystems, and anticipated advances in technology will help us address remaining challenges. Among the issues needing attention are the declines in populations of aquatic species that are not only environmentally essential but also economically vital, and "non-point" sources of pollution — that is, pollution that arises from wide areas — such as nitrogen and phosphorus runoff from agricultural fields or oil and sediment from urban development sites.

Standards that ensure that the nation's public water supplies remain safe for human consumption have helped prevent 200,000 to 470,000 cases of gastrointestinal illnesses each year. The Environmental Protection Agency is also working with the states and other stakeholders to develop long-term protection programs, an effort that has led to implementation of special protection programs in about 4,000 communities across the country.

Surveys of the nation's largest rivers show that the number of rivers, lakes, and estuaries safe enough for fishing and swimming has increased by 20 percent. Clean water is essential not only for health reasons, but also for direct economic benefit from fishing, tourism, and other water-based commercial activities.

### Preventing Future Harm

The greater scientific understanding of the environment has enabled us to shift from the initial environmental focus of cleaning up major "point sources" of pollution to a new generation of environmental tools that emphasizes pollution prevention. Sustainability requires that future economic growth be achieved without unacceptable levels of pollution or unsustainable rates of resource use. And science is providing the analytic tools for policymakers and decisionmakers to understand — in advance — the environmental consequences of alternative management strategies. The technologies that help us observe, compute, and communicate will increasingly allow us to anticipate environmental issues in a much more timely fashion. For example, as we further refine computer modeling, we will be better able to simulate interactions among biological, chemical, and physical forces and phenomena to predict a range of outcomes, providing better documentation for policy decisions.

Continuing a comprehensive program of environmental research and development will improve our ability to prevent problems in the future. Federal funding for environmental science provides the technical basis for sound environmental policies that enable us to continue to create jobs and expand our economy without sacrificing human health or healthy ecosystems on which human prosperity ultimately depends.

## Partnerships for a Cleaner Environment

Partnerships among government, industry, and educational institutions can generate new technologies that will grow our economy and help our environment at the same time. The Federal government has taken a leadership role in initiating partnerships designed to fulfill all of these objectives.

The Partnership for Advancing Technologies in Housing (PATH), for example, links key agencies in the Federal government with leaders from the home building, product manufacturing, insurance, financial and regulatory communities in a unique partnership focused on technological innovation in the American housing industry. The goals of the partnership include cutting the environmental impact and energy use of new housing by 50 percent or more. The five PATH National Pilot projects are Village Green and Playa Vista, in Los Angeles; Civano, in Tucson; Stapleton Airport, in Denver; and Summerset, in Pittsburgh.

These programs serve as models for the U.S. construction and housing industry because of their new approaches to land planning and design and their incorporation of highly innovative technologies.

In the Partnership for a New Generation of Vehicles (PNGV), different sectors are combining forces to unlock new technologies that will develop a new class of vehicles with a fuel efficiency of up to 80 miles per gallon and maintain performance, safety, and cost comparable to today's cars. PNGV joins seven Federal agencies and 19 Federal laboratories with the U.S. Council for Automotive Research (USCAR), which represents Daimler-Chrysler, Ford, and General Motors. The PNGV partnership ultimately will help create new jobs, improve global competitiveness, reduce U.S. dependence on foreign oil, and decrease greenhouse gas emissions.

## Chapter 6

# Manufacturing Technologies

Manufacturing undergirds our nation's economy. Manufacturing firms consistently generate about 20 percent of GDP and employ about 16 percent of the total workforce, or about 21 million people. Continual innovations in manufacturing technologies sustain the vital economic role of manufacturing industries in the U.S. economy.

Three decades ago, U.S. manufacturing was concentrated in large factories using large amounts of raw materials to produce machinery, automotive vehicles, and other large products. Labor was skilled but relatively expensive to the manufacturer, who often had to tread a fine line between cost and quality concerns. Today's manufacturing model is a much smaller factory producing smaller consumer goods or precision parts for later assembly in larger products. Miniaturization, new materials, and improved processes have helped manufacturers make great strides in quality, efficiency, and productivity. This rapid rate of progress is fueled by research in manufacturing systems, as well as innovations in a range of other disciplines — including materials science, robotics, chemistry, information technologies, management, and statistics.

### Influence of Information Technologies

Looking at just one of these contributing disciplines, we can see that information technology has been a powerful driver of innovation in manufacturing. Design and manufacture of new products requires processing of huge amounts of information to ensure quality, satisfy customer needs, and meet environmental standards. Increasingly, manufacturers are using information technology for modeling and simulation to construct "virtual" tools and factories, allowing factory design to proceed in parallel with product design. Similarly, computer networking helps manufacturers integrate all aspects of their operations, from design and processing to the assembly line, shipping, and marketing. These applications of information technology reduce the time and cost of developing and testing new products and allow them to reach their markets faster and more efficiently.

Manufacturers are increasingly incorporating computers and other forms of technology into the workplace, both on the factory floor and on the office desktop. This has increased productivity, which in turn, has boosted worker compensation. Since the fourth quarter of 1995, nonfarm business productivity growth has averaged 2.1 percent. Nowadays it is difficult to identify a "low-tech" manufacturer: 84 percent of manufacturers use computer-aided design

## Pollution Prevention Pays

A program at 3M to encourage innovation among employees has not only helped the company improve speed and efficiency, but has also helped create a cleaner environment and generate new revenues. The Pollution Prevention Pays (3P) program at 3M aims to prevent waste at its source — in products and manufacturing processes — rather than treating or disposing of it after it has been created. Although the idea itself was not new when 3P began in 1975, no one had ever tried to apply pollution prevention on a company-wide basis and document the results. Since 1975, 3P has kept 771,000 tons of pollutants out of the environment and saved $810 million.

Before the 3P program, a resin spray booth in one plant had annually produced about 500,000 pounds of over-spray, requiring special incineration disposal. The company installed new equipment to eliminate excessive over-spray. It also implemented a new design that reduced the amount of resin used. In this case, an equipment investment of $45,000 saved more than $125,000 a year.

Another 3M plant developed a new product from the waste stream of an existing product at the plant. The new product is used to contain and absorb hazardous waste spills, providing revenue, cutting landfill costs, and reducing waste. Other 3P projects worldwide have ranged from improved control of coating weight and wastewater recycling, to a variety of combustion control and heat-recovery processes.

*The Federal investment in environmental research is helping to encourage American corporations to develop manufacturing processes to minimize pollution.*

(CAD); 63 percent have incorporated local area networks (LANs) into their operations; and 62 percent have adopted "just-in-time" inventory techniques.

Before a company starts full-scale manufacturing of a product, it builds prototypes with the same specifications as those of the planned product. The prototypes are used for testing and verification of the design and error-proofing manufacturing assembly. Older methods for constructing individual prototypes were expensive and time-consuming, adding substantially to the time between product concept and delivery. Today, rapid prototyping reduces prototype development time from months to days, greatly shortening the time to market of new products.

In addition, increasing productivity and capturing a world market depend on "agility" in manufacturing — setting up manufacturing enterprises to adapt products rapidly to changing marketing opportunities in the most efficient way. Companies are finding ways to perfect "just-in-time" procedures in assembly, inventory, and delivery, so that resources — including human effort — are applied when they are needed, but not until then. This approach requires rapid flow and application of information roughout the supply chain. The end result is substantial increases in productivity, as reflected in savings of money and manpower throughout the manufacturing sector.

*continued on page 28*

## Virtual Manufacturing

Designing, testing, and developing large manufactured products requires many human and material resources. Information technologies help integrate computer design tools with models and simulations of manufacturing processes for more efficient design, analysis, and testing of products. These 'virtual' tools greatly reduce the investment required for product prototyping, testing, and validation. The story of the development and production of the Boeing 777 is a vivid illustration of the adoption of virtual manufacturing and the efficiencies that technology can create.

The latest relative in Boeing's family of jetliners, the 777 is the first airplane to be completely designed and pre-assembled 'virtually' — that is, by computer. Performance and strength of the plane were analyzed and tested through complex computer models. Of its three million parts, more than 100,000 are unique; they were precision-engineered from computer models. The parts were manufactured separately at sites spread around the world, then shipped to a central plant, where they were assembled. They fit together perfectly on the first attempt! The cost savings to Boeing were tremendous, and the company won multiple manufacturing and innovation awards.

*Our ability to harness the power and promise of leading-edge advances in technology will determine in large measure our national prosperity, security, and global influence, and with them the standard of living and quality of life of our people.*

## Customizing Mass Production

U.S. manufacturing firms are adopting techniques that are potentially as revolutionary as Henry Ford's development of the Model T automobile. The mass-produced auto epitomized the industrial revolution; the assembly line standardized quality, reduced costs, and passed on these benefits to the consumer. But the consumer also had to accept fewer choices — most famously, every Model T was painted black.

Today, the advent of information technology is changing the nature of manufacturing and raising consumer expectations. In a world where we have grown accustomed to instant Internet access to specific information on almost any topic, we increasingly expect products to be tailored to our individual needs. Already, customers are using the World Wide Web to configure their dream car or their next computer. With a click, their order goes directly to the manufacturing plant.

Even more sweeping are IT-enabled changes in manufacturing practices and business relationships. Supply chains span the globe, linked in information-sharing networks that rapidly exchange designs, part orders, demand forecasts, sales reports, and much, much more. Without leaving their home offices, equipment manufacturers can go on line to troubleshoot — and even correct — problems in a customer's plant hundreds of miles away, saving time and money. A small manufacturer with occasional need for a costly design or research tool can contract, via the Internet, with a specialized service provider, bringing the company the benefit of unique expertise without having to hire new staff. And, in the steel industry, companies are trimming storage costs by advertising and selling surplus production via their Web sites.

Some companies already are making customized products on production lines that are only a link or two away from the customer. Dell Computer Corporation, for example, uses a computerized system that informs workers which components to install, according to customer specifications on orders received on the company's Web site. The system automatically reorders components according to demand, a practice that reduces surplus inventory and prevents volatile components from losing value (up to 1 percent per week). This system works well for building computers, whose parts can be configured in many different ways according to customers' needs, but many other industries also use the technology. In the apparel industry, some companies are scaling production runs to orders as small as one item. Their customers supply measurements over the Internet, and the firms send back attire that truly is made to fit.

Economists credit applications of information technology for driving annual productivity increases in manufacturing that have been averaging about 4 percent since 1992, double the rate of increase for other non-farm sectors of the economy. Manufacturers are still finding new, productive uses of information technology. In the decades to come, information technology will bring the Industrial Revolution full circle, and mass-produced customized products will become the norm.

*Next generation vehicles such as this one will incorporate advances in manufacturing and information technologies.*

*The Federal government forms partnerships with companies to develop technologies to achieve broad-based economic benefits with high rates of social return for the nation.*

## Small Component, Big Impact

The health of the U.S. printed wiring board industry has improved dramatically in the past several years, thanks to a collaborative research venture co-funded by the Advanced Technology Program (ATP) of the Department of Commerce.

Printed wiring boards are a powerful but unseen component of our modern Information Age — in fact, most people have never seen one. Nonetheless, they are crucial in the operation of dozens of products we use every day, from copy machines, pagers, and computers to radar, industrial sensors, and biomedical implants. These wiring boards connect smaller electronic devices inside the products. Between the early 1980s and the early 1990s, the $7 billion industry, which represents some 200,000 American jobs, was steadily losing world market share. Then the ATP partnered with six top U.S. suppliers and users of printed wiring boards and Sandia National Laboratories of the U.S. Department of Energy to look for ways to improve the industry's manufacturing efficiency.

Between mid-1991 and mid-1996, the venture hastened progress and substantially reduced the costs of 32 research tasks and enabled the industry to pursue 30 other tasks that would not have been possible without ATP funding. The initial gains in productivity were remarkable: one company reduced the number of plies, or layers of material, in its wiring boards, saving more than $3 million annually; another company used a new model for predicting shrinkage of its wiring boards' layers, reducing its accumulation of scrap and saving more than $1.4 million per year; and a third firm found ways to improve its coating and soldering techniques, reducing solder joint defects by 50 percent. The venture succeeded not only through technical accomplishments but also through spin-off projects that may further boost the industry's fortunes — especially in the dynamic market for portable electronics. One group of engineers involved in the project started a new company that now tests sample boards for major corporate clients around the world.

The industry saved a total of $35.5 million in research costs, and millions more via increased productivity. One expert credits the ATP program with saving the entire U.S. industry. The U.S. share of the market for printed wiring boards has increased from a low of 26 percent in the early 1990s to 31 percent in 1996, and orders were up nearly 20 percent as of mid-1997. Ultimately, the biggest beneficiaries of the reduced costs and improved quality in these products are American consumers.

*continued from page 24*

## Federal Support for R&D

For several decades, the Federal government has supported research relevant to various manufacturing sectors. The government also provides economic and technical information to manufacturers, and it establishes and nurtures partnerships and consortiums involving universities, private manufacturers, and broader industry groups. For example, the Commerce Department's National Institute of Standards and Technology (NIST) is a key supplier of technologies and services integral to manufacturing capabilities. Results of NIST research lead to industry-accepted test and measurement methods, process models, interface standards, and other useful tools that contribute to effective operations and quality products across a wide range of manufacturing industries. The capabilities that they support often set the technical limits on what can be accomplished on the factory floor, in the research and development laboratory, or with suppliers and customers. American companies depend on NIST tools and services for hundreds of millions of measurements each day.

## Nanotechnology and Beyond

Today's research provides a glimpse of the future of manufacturing. Scientists are now able to see things at the molecular level, and are rapidly gaining the ability to manipulate materials and processes at the nano-level. (A nanometer is one-billionth of a meter, tens of thousands of times smaller than the width of a human hair.) In the emerging field of nanotechnology, researchers are working to find ways to change the very composition of materials to emphasize desired characteristics such as strength or flexibility. Nanotechnology holds tremendous promise for the future of manufacturing, signaling a new ability to custom-design materials that manufacturers might need for specific purposes. Within an estimated one to three decades, nanofabrication processes will move from the laboratory to the assembly line, and new nano-materials will find countless new applications in products and processes that will achieve even greater efficiencies and quality levels.

We can even expect to see molecular-size switches for computer circuits; "machines" no bigger than a few atoms; surgical tools that can operate on an individual cell in the human body; and molecular robots that doctors can inject into the bloodstream, where they will seek out and destroy cancer cells. Nano-engineers are already envisioning "self-assembling" devices that will rebuild copies of themselves, molecule by molecule, following programmed instructions.

Today, U.S.-based manufacturing extends its global market share leadership mostly through high-tech exports such as computers, semiconductors, software, aircraft, pharmaceuticals, biotechnology, on-line services, telecommunications, and precision instruments. We are beginning to transform manufacturing processes and equipment by intelligent sensors and control systems, rapid prototyping capabilities, and pollution avoidance technologies. America's leadership in manufacturing has not been without global challenges, but we are witnessing a surge of innovation that will enhance our nation's global economic manufacturing capability.

As newly developing fields such as nanotechnology, fiber optics, robotics, and computer modeling continue to yield breakthroughs in products and processes, we will undoubtedly see even more dramatic changes in the coming century. Even if what is being manufactured is the same 30 years from now, it's a safe bet that how it's manufactured will be cleaner, more efficient, and more productive.

# Epilogue

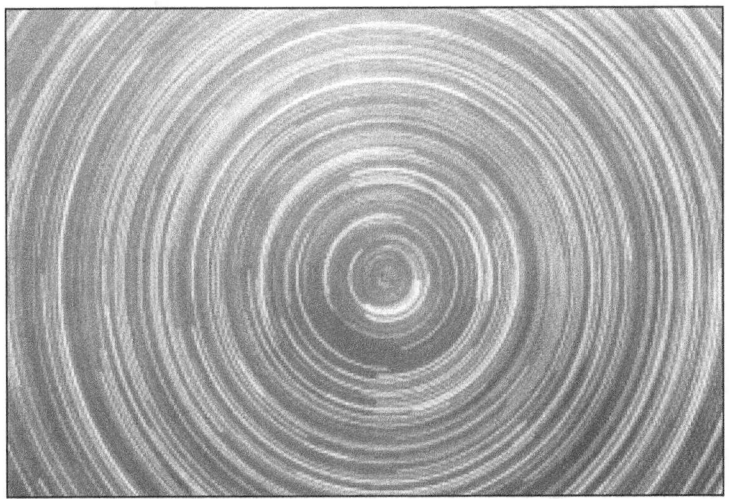

Wherever we turn in our daily lives, we constantly encounter reminders of the contributions of science and technology. From the familiar (for example, a phone call via optical fiber cable) to the astonishing (the successful cloning of Dolly the sheep), examples abound of technology's pervasiveness. As a society, we will wrestle with moral and ethical questions raised by some of the newest capabilities we have developed, such as stem cell research or genetically modified foods. But we should remember that much of what we now take for granted as gifts from science and technology could not have been foreseen decades earlier, and would not have been available without vigilance in research and development funding.

If we are to continue to enjoy beneficial breakthroughs from scientists and researchers, we must make sure that we continue to fund essential research and development activities across a broad spectrum of scientific disciplines. Only by supporting research where the returns are not guaranteed can we ensure the steady, gradual progress that underpins the front-page news stories that accompany each new success. It is ironic that such open-ended research, whose cost-effectiveness is often difficult to guarantee, sometimes generates the greatest economic returns. Federal Reserve Chairman Alan Greenspan expressed this very point in the summer of 1999 when he said, "The evidence…for a technology-driven rise in the prospective rate of return on new capital, and an associated acceleration in labor productivity, is compelling, if not conclusive." The President's Committee of Advisors on Science and Technology, in issuing this report, urges all Americans — from Capitol Hill to Main Street — to do all they can to support continued Federal funding for science and technology, so that our grandchildren can continue to benefit from the same wellspring of prosperity.

## About the President's Committee of Advisors on Science and Technology

President Clinton established the President's Committee of Advisors on Science and Technology (PCAST) by Executive Order 12882 in November 1993. The responsibilities of PCAST are "to advise the President on issues involving science and technology and their roles in achieving national goals, and to assist the National Science and Technology Council (NSTC) in securing private sector participation in its activities." NSTC is a cabinet-level council chaired by the President that coordinates research and development policies and activities across Federal agencies. The formal link between PCAST and NSTC ensures that the private sector perspective is included in the policy-making process.

### Members

Neal F. Lane, *Assistant to the President for Science and Technology, and Director of the Office of Science and Technology Policy* (**Co-chair**)

John A. Young, *Former President and CEO, Hewlett-Packard Co.* (**Co-chair**)

Norman R. Augustine, *Former Chairman and CEO, Lockheed Martin Corporation*

Francisco J. Ayala, *Donald Bren Professor of Biological Sciences, Professor of Philosophy, University of California-Irvine*

John M. Deutch, *Institute Professor, Department of Chemistry, Massachusetts Institute of Technology*

Murray Gell-Mann, *Professor, Santa Fe Institute; R.A. Millikan Professor Emeritus of Theoretical Physics, California Institute of Technology*

David A. Hamburg, *President Emeritus, Carnegie Foundation of New York*

John P. Holdren, *Teresa and John Heinz Professor of Environmental Policy, John F. Kennedy School of Government, Harvard University*

Diana MacArthur, *Chair and CEO, Dynamac Corporation*

Shirley M. Malcom, *Head, Directorate for Education and Human Resources Programs, American Association for the Advancement of Science*

Mario J. Molina, *Institute Professor, Department of Earth, Atmospheric and Planetary Sciences, Massachusetts Institute of Technology*

Peter H. Raven, *Director, Missouri Botanical Garden; Engelmann Professor of Botany, Washington University in St. Louis*

Sally K. Ride, *Professor of Physics, University of California-San Diego*

Judith Rodin, *President, University of Pennsylvania*

Charles A. Sanders, *Former Chairman, Glaxo-Wellcome Inc.*

David E. Shaw, *Chairman, D.E. Shaw and Co. and Juno Online Services*

Charles M. Vest, *President, Massachusetts Institute of Technology*

Virginia V. Weldon, *Center for the Study of American Business, Washington University in St. Louis*

Lilian Shiao-Yen Wu, *Research Scientist and Consultant, Corporate Technical Strategy Development, IBM*

### For More Information

President's Committee of Advisors on
Science and Technology
1600 Pennsylvania Avenue, NW
Washington, DC 20502
202.456.6100

information@ostp.eop.gov
http://library.whitehouse.gov/WH/EOP/OSTP/NSTC/PCAST/pcast.html

www.ingramcontent.com/pod-product-compliance
Lightning Source LLC
Chambersburg PA
CBHW081813170526
45167CB00008B/3422